discovermore

Exploring Life Science

Understanding Animal Structures

Ernest Brazzos

Britannica Educational Publishing

IN ASSOCIATION WITH

THE ROSEN PUBLISHING GROUP

Published in 2025 by The Rosen Publishing Group, Inc.
2544 Clinton Street, Buffalo, NY 14224

Editor: Brianna Propis
Book Design: Michael Flynn

Photo Credits: Cover Krasula/Shutterstock.com; (series background) Dai Yim/Shutterstock.com; p. 4 Valerijs Novickis/Shutterstock.com; p. 5 Gialdini Luca/Shutterstock.com; p. 6 Gerald Robert Fischer/Shutterstock.com; p. 7 Garna Zarina/Shutterstock.com; p. 8 MintImage/Shutterstock.com; p. 9 David A Litman/Shutterstock.com; p. 10 Natalia Saudi/Shutterstock.com; p. 11 Alexandra Giese/Shutterstock.com; p. 12 Eric Isselee/Shutterstock.com; p. 13 (top) AgriTech/Shutterstock.com; p. 13 (bottom) papkin/Shutterstock.com; p. 14 Adam Ke/Shutterstock.com; p. 15 Ethan Daniels/Shutterstock.com; p. 16 Martin Prochazkacz/Shutterstock.com; p. 17 Jsegalexplore/Shutterstock.com; p. 18 FJAH/Shutterstock.com; p. 19 Dreamcreation/Shutterstock.com; p. 20 Gaschwald/Shutterstock.com; p. 21 Ken Griffiths/Shutterstock.com; p. 22 frantic00/Shutterstock.com; p. 23 Roberto Dani/Shutterstock.com; p. 25 (top) Manop Boonpeng/Shutterstock.com; p. 25 (bottom) Mathias Sehnke/Shutterstock.com; p. 26 Wildpix productions/Shutterstock.com; p. 27 Vecton/Shutterstock.com; p. 28 Thorsten Spoerlein/Shutterstock.com; p. 29 Willyam Bradberry/Shutterstock.com.

Library of Congress Cataloging-in-Publication Data

Names: Brazzos, Ernest, author.
Title: Understanding animal structures / Ernest Brazzos.
Description: Buffalo : Britannica Educational Publishing, [2025] | Series: Discover more! Exploring life science | Includes index.
Identifiers: LCCN 2023051659 (print) | LCCN 2023051660 (ebook) | ISBN 9781641901710 (library binding) | ISBN 9781641901703 (paperback) | ISBN 9781641901727 (ebook)
Subjects: LCSH: Animals--Juvenile literature. | Physiology--Juvenile literature.
Classification: LCC QL49 .B6697 2025 (print) | LCC QL49 (ebook) | DDC 590--dc23/eng/20231129
LC record available at https://lccn.loc.gov/2023051659
LC ebook record available at https://lccn.loc.gov/2023051660

Manufactured in the United States of America

CPSIA Compliance Information: Batch #CSBRIT25. For further information contact Rosen Publishing at 1-800-237-9932.

Contents

Animals

Animals exist in all sorts of shapes and sizes. An orangutan is very different from a dolphin, and a dog is very different from a spider. Yet, they are alike because they have features in common. All animals are living things that need to consume energy, move around, and react to their environments.

Animals consume energy by eating food. Some do this by hunting and killing other animals.

The coconut octopus hides itself using camouflage. Camouflage is when an animal can blend into its surroundings.

Scientists categorize animals by the structure of their bodies. Most animals have no backbone. These are invertebrates. Some animals, including people, have backbones. They are vertebrates.

Many structural features help animals survive. These are called adaptations. For example, the coconut octopus can change its color to hide from predators.

Consider This

Animals and plants are both living things, but they look very different from each other. How might they be similar?

Soft Invertebrates

Worms have no backbone. They are soft, long invertebrates. Some worms are fat, and some are very thin, like threads. Others are flat, like ribbons. Some worms are so tiny you can hardly see them, while others grow up to 100 feet (30 m) long! Worms can survive all over the world, except where the ground is too frozen or dry for them to live.

Flatworms have soft, flattened bodies. Some that live in the sea are brightly colored!

Each earthworm has both male and female reproductive organs.

To eat, an earthworm uses a tongue-like body part to grab food and pull it back into its mouth. They sift out food from the soil and dump the waste dirt.

Some worms feed off other animals. A tapeworm will live inside the stomach and intestines of other animals and eat the food that the animal eats.

Consider This

Worms have no arms or legs yet are able to move around. What adaptations do you think they developed in order to move?

Shelled Invertebrates

Like worms, mollusks are invertebrates with soft bodies. But a mollusk's body has a covering called a mantle. In most kinds of mollusks, the mantle makes a shell. Some mollusks, such as snails, have a single, solid shell. Other mollusks, called bivalves, have two-halved shells that help them swim around and hide from predators.

The largest bivalve is the giant clam. It can grow to more than 4 feet (1.2 m) and can live to 100 years or more!

Cuttlefish have tentacles for catching prey and arms for handling food. You can see its tentacles here!

Certain mollusks, such as octopuses, cuttlefish, and squid, don't have a visible shell. They're called cephalopods.

Consider This

Why don't scientists consider worms to be mollusks?

Underwater Invertebrates

Crustaceans are a very diverse group of invertebrates that include both lobsters and barnacles. Most of these invertebrates have a body protected by a tough **exoskeleton**. Crustaceans have segmented bodies like insects. This means their bodies are divided into three sections: a head, a middle part called a thorax, and an abdomen.

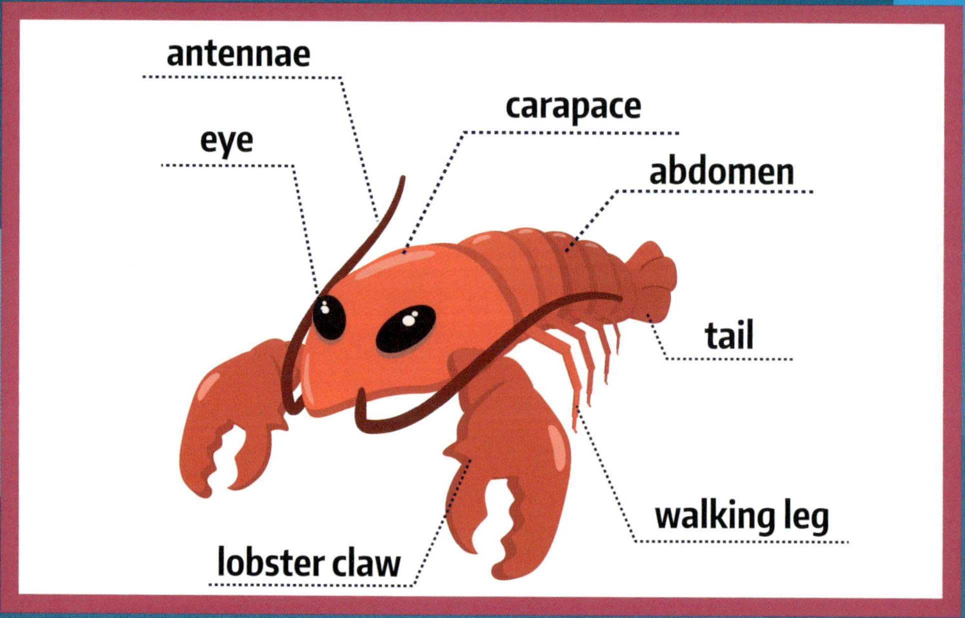

This illustration shows the many parts of a lobster's body.

Crustaceans are members of a group of creatures called arthropods. Arthropod is a Greek word meaning "jointed leg."

Crustaceans have 10 legs, some of which can be claws or swimming paddles. Some have strong claws on their front legs to pick up food and to defend, or protect, themselves. They use their other legs to scuttle over rocks or to swim. Barnacles are unusual crustaceans that are stuck to rocks. They wave their legs to catch food from the water. Crustaceans live mostly in water, including both fresh water and salt water. Some can live on land, such as hermit crabs and woodlice.

WORD WISE

AN EXOSKELETON IS A HARD COVERING THAT SUPPORTS AND PROTECTS THE BODIES OF CERTAIN ANIMALS.

Bugs

Like crustaceans, insects and spiders have an exoskeleton. However, there are many differences between their structures. All insects have three main body segments. They have a head, which often has large eyes. They have a thorax with six legs. They have an end part, called the abdomen. Although not all insects have wings, the ones that do are the only invertebrates with wings.

Spiders have two main body segments and eight walking legs. No spider has wings. Spiders often make webs to catch prey.

Scorpions are close relatives of spiders, but they don't spin webs. Instead, they quickly grab prey with their pincers and sting them with the venomous tips of their tails.

This is a close-up photo of a scorpion's deadly tail!

Insects have two pairs of wings that are part of their exoskeleton.

compare and contrast

How are crustaceans and insects similar? How are they different?

Spiders spin webs from silk that's made inside their bodies.

Spiny Skin

Invertebrates that have a spiny skin are called echinoderms, which include starfish, sea urchins, and sea cucumbers. An echinoderm's body has equal sections arranged around a central point. Many starfish have five arms. Urchins have five equally sized parts spreading out from their centers.

Although "fish" is in their names, starfish aren't fish. Many people call them sea stars!

If a sea cucumber is attacked, they will release their organs from their mouth!

Echinoderms' skeletons are on the inside of their bodies, which is called an endoskeleton. Echinoderms have tubes filled with water in their bodies. These are connected to rows of tiny tube feet under their bodies. Squeezing water through the tubes makes the feet suck the ground and moves the animals along.

Consider This

If a starfish's arm or leg is broken off, the creature can grow a new one! Can you think of any other animals that can regrow their limbs?

Fins and Tails

Fish can look as different as pufferfish and sharks, or goldfish and eels. These vertebrates have gills to breathe underwater. They use fins and tails to swim.

Sharks and rays have a skeleton made from cartilage. Your nose is made from this bendable material, too. These fish are covered in V-shape scales, like teeth, that allow them to swim faster.

The great white shark is a large, powerful fish. They have 300 sharp teeth that help them eat their prey!

Tuna fish are a type of bony fish. They are also a popular fish for people to eat.

Some fish have hard, bony skeletons inside their bodies. Bony fish have smooth, overlapping scales on their skin. Their gills are covered by flaps on the sides of their bodies.

Consider This

What structures do fish have that help them live their lives underwater?

Slimy Skin

Amphibians are vertebrates that live both in water and on land, and they often have slimy skin. These include frogs, toads, and salamanders.

Most amphibians lay jelly-like eggs in water. The young that hatch have gills and swim around like fish. As they grow, amphibians develop legs and breathe using lungs.

A frog's bulging eyes allow it to see in front, on its sides, and somewhat behind it.

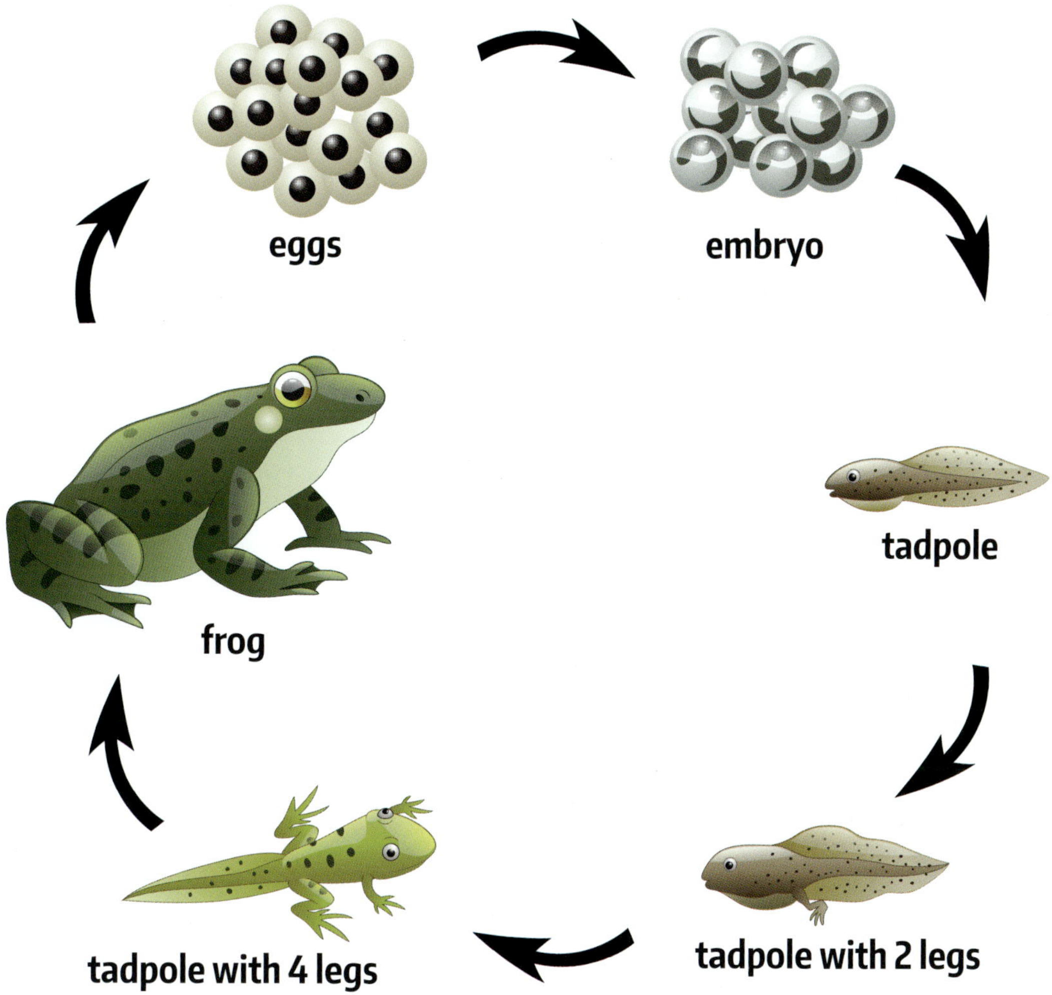

This drawing shows a frog's life cycle from an egg to a tadpole to an adult.

The largest group of amphibians are frogs and toads. They have large heads, short bodies, no tail, and long back legs. A frog's toes are webbed so they can swim.

Scaly Skin

Vertebrates with dry, scaly skin are called reptiles. These include lizards, snakes, turtles, alligators, and crocodiles. They breathe using lungs. Most reptiles lay eggs. Certain snakes and lizards give birth to live young.

Lizards include iguanas and geckos. They have long tails and overlapping scales. Lizards are similar to crocodiles and alligators. However, a lizard's scales overlap, and they have bony eye rings unlike crocodiles and alligators.

A chameleon can change its skin to blend in with its surroundings. This is called camouflage.

Snakes smell with their tongues. They flick their tongues to gather scents from the air.

Some reptiles have shells covering their backbone. This includes turtles, which live mostly in water, and tortoises, which live on land.

Snakes are legless reptiles with no eyelids. They can be as enormous as anacondas or as tiny as threadsnakes.

Consider This

Why might a chameleon need to camouflage itself?

Feathers and Wings

All birds have wings, feathers, and a light skeleton. But they can look very different from one another, like an ostrich and a sparrow or a swan and a blue jay. Their bodies have adaptations for flight. Some birds, such as penguins, have wings but can't fly. They use their wings to swim. All birds are a part of the vertebrates group.

Birds developed beaks so they could eat their food. Scientists believe birds descended from dinosaurs!

Among the 10,000 bird species in the world, five species can't fly or sing: the ostrich, emu, rhea, kiwi (pictured), and cassowary.

A bird's beak is adapted to the food they eat, so different birds have different beaks. For example, a hawk's bill can rip apart prey.

Birds have different feet, too. Some have claws and curved toes to perch on branches. Others have webbed feet so they can swim on or under the water.

Milk and Fur

Mammals are vertebrates with hair or fur on their bodies. This keeps them warm and dry. Human beings are mammals.

Female mammals feed their young with milk from their body. Baby mammals usually develop inside their mothers' body. Some animals, such as kangaroos, have pouches on their fronts in which their babies develop. Certain mammals even lay eggs, such as platypuses.

Although they have things in common, a mammal's body will be different—depending on where it lives—because of adaptation. Whales and dolphins have smooth bodies, broad tails, and flippers to swim underwater. Gibbons, a type of ape, have long arms to swing through trees. The Arctic fox has thick fur that changes colors in winter and summer to match its surroundings.

A mother's milk helps her baby to grow and stay healthy.

compare and contrast

Birds have different beaks depending on the food they eat. What might be different about various mammal species, such as deer and wolves, to help them eat different foods?

Mammals breathe using their lungs. Dolphins and whales come to the surface for air because they don't have gills and can't breathe underwater like fish.

Classifying Animals

Scientists group animals into categories so it's easier to notice their similarities and differences. This is called classification. We classify animals partly based on structures, such as wings. We also classify by behavior, such as whether they eat other animals.

It may be difficult to classify some animals. Horseshoe crabs look like crabs, but they're more closely related to scorpions. The platypus is a mammal with fur like an otter, but has the bill of a duck, the tail of a beaver, and webbed feet. It also lays eggs!

Although it may look like a mouse, squirrel, or even a tiny kangaroo, a gerbil is a member of the family of jumping rodents.

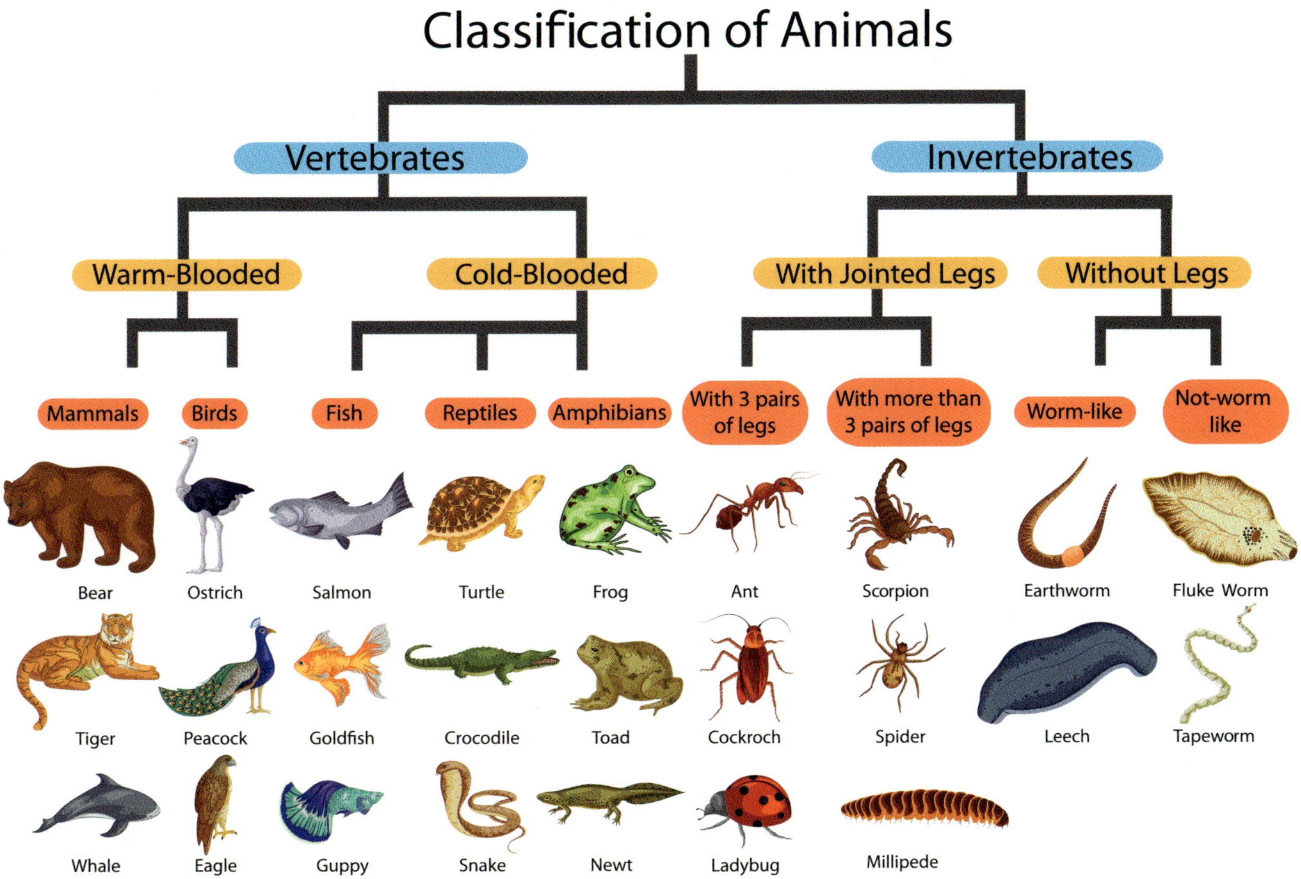

This illustration shows the classification of vertebrates and invertebrates and breaks down the two groups even further!

Adaptation

Animals within the same species will often look different depending on their sex. Male peacocks have colorful tails, but females are brown. A male mallard duck usually has a bright yellow bill, and the female has an orange and brown bill.

Can you tell which mallard is male and which one is female?

Plastic that ends up in the ocean is often consumed by sea turtles, which is deadly to them.

Changes in nature can also have an effect on an animal's features. A dog will grow thicker fur during colder months and will lose more fur in warmer months.

Sometimes people make changes that affect animals. When humans **pollute** rivers and ponds, frogs and salamanders can be killed, their immune systems may be weakened, or they may even grow extra legs.

WORD WISE

POLLUTION IS ANYTHING THAT MAKES NATURE DIRTY AND UNHEALTHY. IT IS OFTEN CAUSED BY HUMAN ACTIVITY.

Glossary

abdomen The rear section of an insect's three body segments.
carapace The hard upper shell of a turtle or crustacean.
cartilage A tough, flexible material supporting the body of some fish. Cartilage is also found in other animals.
descend To come down from a certain group or family.
diverse Showing a lot of variety.
environment Everything surrounding both living and nonliving things.
hermit crab A type of crab that lives in the empty shell of another animal, such as a snail.
orangutan A large red ape with long reddish hair and long arms.
pincer A claw-like body part on an insect that helps it to grab food.
pufferfish A type of fish with spiny skin. It can inflate itself like a balloon when it feels threatened.
rodent A type of mammal with a special kind of teeth. A rodent can be a mouse, hamster, beaver, or squirrel.
shrimp A small shellfish.
species A group of similar organisms that can reproduce, or make offspring.
woodlice A flat, gray crustacean that lives in damp places, like rotting logs or fallen leaves.

For More Information

Books

Daniels, Patricia. *Mammals.* Washington, D.C.: National Geographic, 2019.

Earley, Christina. *Classifying Animals.* Coral Springs, FL: Seahorse Publishing, 2023.

Krekelberg, Alyssa. *Essential Invertebrates.* Minneapolis, MN: Essential Library, 2022.

Websites

Britannica Kids: Adaptation
www.kids.britannica.com/kids/article/adaptation/352704
Learn more about the ways animals have adapted to their surroundings for survival.

National Geographic: Plant and Animal Reproduction
www.education.nationalgeographic.org/resource/plant-and-animal-reproduction
Explore the ways plants and animals pass on their genes.

National Geographic: Vertebrate or Invertebrate
www.education.nationalgeographic.org/resource/vertebrate-invertebrate
Learn more about vertebrates and invertebrates with this helpful graphic.

Index